MAX BRUIN
MARJOLIJN
MAX KISMA
VAN ESSEN
STEWART McD... KEVIN KELLY (US)
STEVEN HELLER (US) AARON BETSKY (NL/US)
JAN VAN DEN BERG (NL) HENK OOSTERLING
(NL) LUNA MAURER (DE) ANNELYS DE VET (NL)
LEV MANOVICH (US) JOACHIM SAUTER (DE)
PAUL FRISSEN (NL) SILKE WAWRO (DE) ERIK
SPIEKERMANN (DE) SUSANNA DULKINYS (US)
MATTHEW FULLER (UK) MICHAEL ERLHOFF
... BOS (NL) STEFAN SAGMEISTER (US)
... ÖM (SE) NIKI GOMEZ (UK)
... (DE) MARTIJN HAZELZET (NL)
... ORN (SE) UTA BRANDES (DE)
ANDREW ... (UK) MANUELA PORCEDDU
(NL) DOUGLAS RUSHKOFF (US) RICHARD JOLY
(CAN) SIGGA SIGURJONSDOTTIR (ICELAND)
ARMIN MEDOSH (UK) SOPHIA DRAKOPOULOU
(UK) JONAH BRUCKER-COHEN (US) MEG
MCLAGAN (US) MARK FITZPATRICK (UK) TARA
KARPINSKI (US) PAUL MIJKSENAAR (NL)
RICHARD BARBROOK (UK) SADIE PLANT (UK)
PATRICK LICHTY (US) WILLEM VAN WEELDEN
(NL) NATACHA VAIRO (CAN) MCKENZIE WARK
(AUS) ANDREU BALIUS (ES) STEFFEN P WALZ
(US) GEERT LOVINK (NL/AU) JANINE HUIZENGA
(NL) ROBIN HAMMAN (UK) SHULEA CHEANG (US)

# WELCOME TO MOBILE-VISUAL CULTURE

Welcome to the mobile world of quotes, essays, statistics and factoids, all reflecting the very young state of the art in wireless thinking. This publication asks what it means to become cellular, think wearable and live mobile. Liberated from cables and heavy objects, the new human condition of always being available is a remarkably light and unreflected one. Mobile phones seem to fit in almost unconsciously our busy everyday lives.

Involvement of artists and designers in the development of both mobile interfaces and content is still in an early stage. The main reason for this could be the proprietary nature of the devices and their software. It is yet uncertain what visual culture the mobile environment will bring. How colorful will the mobile world be? Is it appropriate to make a comparison with the Internet? Usage of mobile phones may exceed that of the Net, but are users and producers of content in control to the same extent?

The contributors to Mobile Minded are pioneering a new critical discourse. There is no mobile phone theory, yet. While you read and browse, the vocabulary is in the making. The question raised here still is one of bewilderment and excitement. What is the mobile condition?

Will visual culture disappear in the future and will we instead, for instance, use our ears to experience beauty and excitement? Now that globalization has brought us worldwide visual inflation, we are closing our eyes and opening up other senses and exploring new parts of the body. The question than becomes: will we get a similar imagery delivered on our mobile devices, this time embedded in our bodies? Does it really make sense to repeat the telephone, radio and television culture of the twentieth century, this time delivered in a matchbox?

The current situation concerning wireless technologies seems to be ambivalent. Whereas the use of mobile phones worldwide seems unstoppable and continues to grow in an unprecedented pace, the involvement of artists seems to grow roother more slowly. It seems hard to go beyond the level of the ordinary consumer.

However, over the past few years we have seen a gradual rise in arts project which use 'cellspace' as their communication environment. Who else are promoting mobile phone projects in order

# BY MIEKE GERRITZEN & GEERT LOVINK

to shape a rich and diverse public space within this highly commercial, and controlled virtual environment? The so far problematic relationship between the more or less free Internet environment and the highly edited information streams accessible via mobile devices is certainly a controversial topic which is not going to be resolved overnight.

Mobile technology has liberated objects from their serfdom. Objects are no longer bound to a determined locality. Instead of staying behind when we go on the move, the techno objects are accompanying us in an almost unconscious manner. Mobile phones express an ambivalent relationship towards locality. While mostly used nearby home and the workplace, cellphones are also part of global information systems.

We carry PDAs and mobile phones of our choice close to our body as if they were our most intimate friends. Often, people don't even get that close, compared to the invasiveness of 'wearable' technologies. They are truly becoming 'extensions of man', as Marshall McLuhan once described media. As cyborgian fetishes for the masses, the tiny electronics navigate us through our busy everyday lives. They assist us in finding the right information ecology (whose call to answer, who to block, which SMS to answer). They help us to beat boredom. Handy phones and portable web browsers reflect the global condition of electronic herds of hyper individual subjects as 'projects' (Vilem Flusser). Always on the move, always accessible, 24/7, on every possible spot you can imagine.

The other side of techno-mobility is the liberating dimension of Becoming Mobile. To operate 'mobile minded' gives us the possibility to freely move around, question authority and predetermined behavior. Freedom of movement is an essential human right and with it comes the possibility to leave behind conservative frameworks which try to stick us to one place and one position, one ideology, thereby preventing people to design their own mindset.

Mobile freedom can only be reached in a 'negative' move in order to defend individual liberty. Liberty means liberty from, and in this context this means the freedom to define ones own technology standards, beyond the phraseology of 'consumer choice.' Technological liberty is a negative. In the act of warding off interference of global telco corporations and their willing government executors, mobile-minded users are shaping their own awareness of digital aesthetics.

# SEND SMS

3337772633_
99966688777_
6444663

**007**

**ONLY in JAPAN**

# WHERE MEN TEND TO VIEW CELLPHONES AS TOYS, WOMAN TREAT THEM LIKE ACCESSORIES

**IMODE: NTT DOCOMO END-USER PRODUCT + HTML INFRASTRUCTURE**

# I DON'T WANT mobile devices

## MY IDEAL IS TO CARRY NOTHING AND TO HAVE THE ENVIRONMENT KEEP UP WITH ME

**EVERY WALL** SHOULD BE A POTENTIAL DISPLAY FOR MY SCREEN

**EVERY GADGET** A POTENTIAL COMMUNICATOR FOR MY NEEDS

**Mobility is for humans** — CARRYING STUFF AROUND IS NOT MOBILITY

**Mobility is for humans** — CARRYING STUFF AROUND IS NOT MOBILITY

ST. HELENS COLLEGE
760
110365
JAN 2004
LIBRARY

008

# LIFE

# FILE

A LEFT-OVER NEWSPAPER SAYS: "MORE THAN 1 BILLION TEXTS ARE DISPATCHED EVERY MONTH, WITH THOUSANDS BELIEVED TO BE DISCARDED EVERY DAY TO AVOID CLOGGING THE NETWORKS."

CUSTOMERS PAY FOR THEIR DELETE AT TEN PENNIES A TIME. THE ARCHITECTURE OF THE NETWORKS HAS TWO BOTTLENECKS: CENTRALISATION; PROFIT. WHICH MEANS SECRECY, WHICH MEANS COMMUNICATION BECOMES SURPLUS TO TELECOMS. BOTH MUST BE DELETED. AT NO COST. TWO FOR THE PRICE OF NONE. THIS BOOK HERE SAYS: "WRITTEN POETRY IS VALID ONCE AND THEN OUGHT TO BE TORN UP."

# I'M A TRAVELER BY PROXY

I'M RINGED. I TRAVEL BY WIRE. WIRES THAT MAKE ME FEEL HOME ABROAD. WIRES THAT BRING FAMILIAR VOICES. WIRES THAT FEED ME CASH. I CAN'T WAIT TO BE WIRELESS. NO LINE FEEDS. NO INTUBATIONS. NOTHING TO HOLD ME BACK. COME TO THINK ABOUT IT, WHY ARE WE STILL BUNDLES OF FIBERS ANYWAY? TISSUE. I WANT THE WAVES. I WANT PURE ENERGY. LIFE. NO STRINGS ATTACHED. NO-ONE TO PULL THEM. I WANT TO TRAVEL BY WAVE. I'D STILL BE TRAVELLING BY PROXY. THE LIGHT WOULD STILL CHANGE, WHEREVER I GO. MY NIGHTS WOULD STILL SMELL OFF ÉGOÏSTE. MY PICS WOULD STILL BE SHOPPED. BUT I WOULD BE FREE.

# UNSTITCHED

# Freedom of SPACE

INSTEAD OF

# Freedom of SPEECH

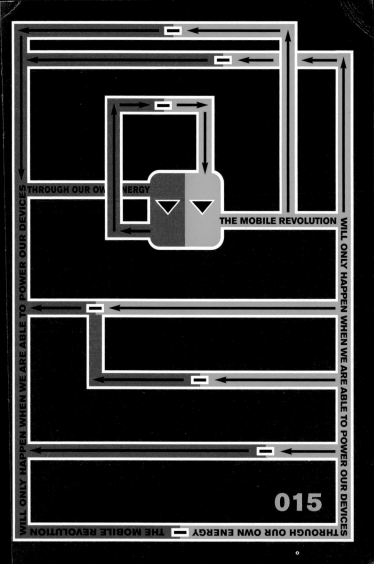

Just as initiation rites once offered access to knowledge and power, so do the media connect us to the ruling culture. Through pseudo-ritual patterns of action, the media are systematically inscribed "INTO OUR FLESH." The more we let ourselves be embraced by the media, the more mediocre we become as users.

# IS MY MOBILE ACTING AS A DOUBLE-AGENT?

018

**BY BECOMING MY EASY-TO-USE GATEWAY FOR ON-LINE BANKING, M-COMMERCE AND SOCIALISING,**

**MY MOBILE IS SURREPTITIOUSLY REVEALING INFORMATION ABOUT MY FINANCES, SHOPPING HABITS AND LIFESTYLE CHOICES TO OUTSIDE FORCES, SUCH AS LAW ENFORCEMENT AGENCIES AND MARKET RESEARCHERS.**

# SPEED

## GSM
**GLOBAL SYSTEM FOR MOBILE COMMUNICATIONS** — UP TO 9600 BAUT

## GPRS
**GENERAL PACKET RADIO SERVICES (2.5G)** — UP TO 114 KBPS

## UMTS
**UNIVERSAL MOBILE TELECOMMUNICATIONS SERVICE (3G)** — UP TO 2 MBPS

## WLL
**WIRELESS LOCAL LOOP** — UP TO 10 MBPS

## WLAN
**WIRELESS LOCAL AREA NETWORK** — UP TO 11 MBPS

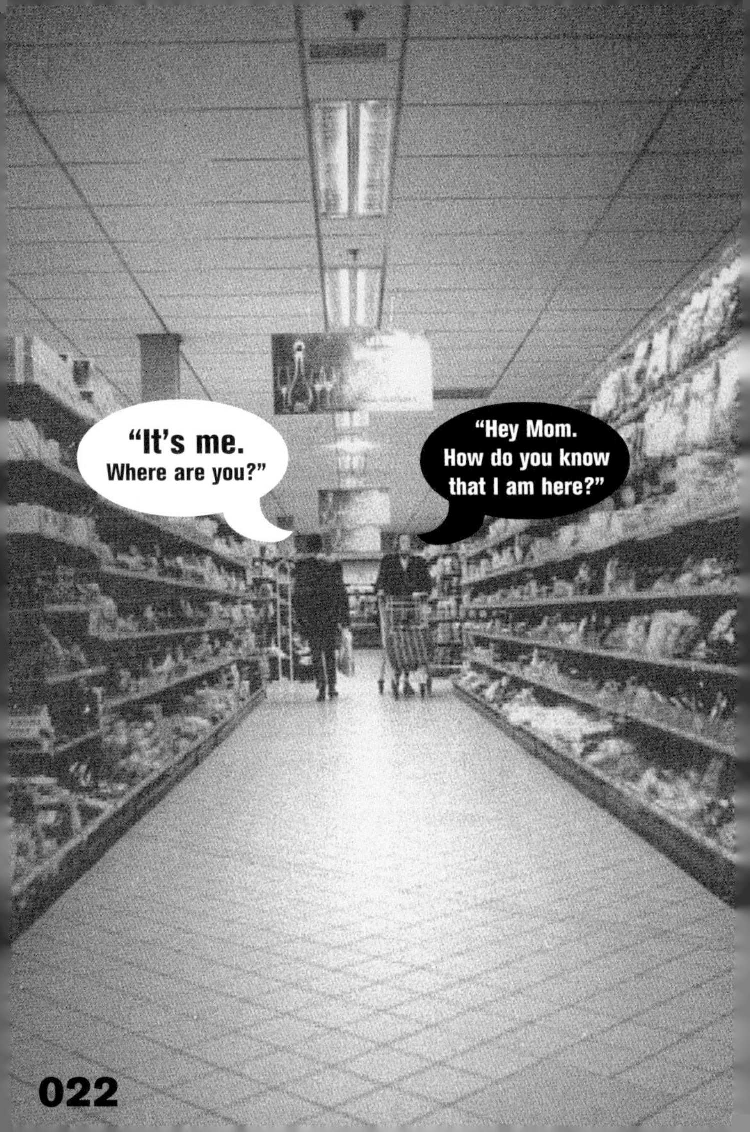
"It's me.
Where are you?"
"Hey Mom.
How do you know
that I am here?"

COMPARED TO CELLPHONES, THE INTERNET WAS NOTHING.

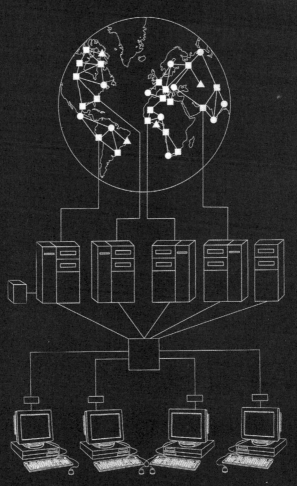

THE INTERNET, FROM DUDEN, BILDLEXICON PAGE 423

**BY GOING MOBILE, YOU BECOME PART OF A WE CULTURE**

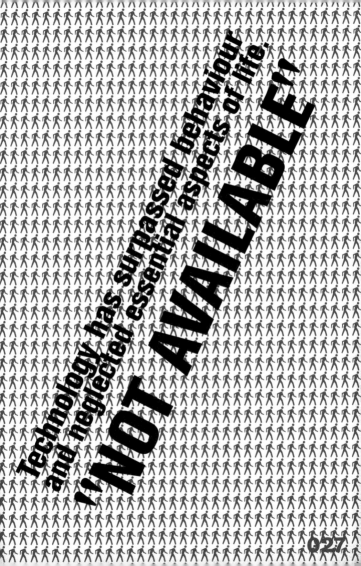

# WHO SAID THAT TEXT WAS DEAD?

**THE POPULARITY OF SMS DISPROVES MCLUHAN'S PREDICTION THAT READING AND WRITING WOULD DISAPPEAR ONCE WE COULD EASILY COMMUNICATE WITH EACH OTHER USING AUDIO-VISUAL MEDIA.**

Gave out soup to people living on the Street in New York yesterday. One of the homeless men in line keeps shouting into his cellphone: FUCK YOU, FUCK YOU, FUCK YOU! When it is his turn in line, he accepts the soup and milk graciously, thanks me nicely for it, and immediately returns to his phone conversation: FUCK YOU, FUCK YOU, FUCK YOU, FUCK YOU!

029

feel H"

you can *feel* visual

# WOULD YOU WEAR THIS SOFTWARE?

you can *feel* sound

030

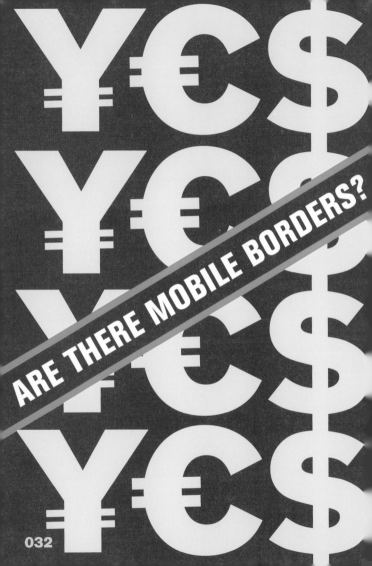

I BELIEVE THAT IN THE GOODNESS OF TIME, MAYBE IN A 100 YEARS FROM NOW, MOBILITY WILL BE CONSIDERED A BASIC HUMAN RIGHT -- EVENTUALLY EDUCATED IN THE UN DECLARATION OF HUMAN RIGHTS.

# THE RIGHT OF MOBILITY
## MEANS THAT
## ANY PERSON ON EARTH CAN
## LIVE ANYWHERE ON EARTH,
## AS LONG AS THEY FOLLOW

★ LOCAL LAWS AND PAY LOCAL TAXES ★

# IMMIGRATION
## LAWS WILL BE
## PERCEIVED AS
## ANTIQUATED AS
## SERVANT AND
## SLAVERY LAWS

# WHO NEEDS MICROSOFT OR LINUX?

# IS A MOBILE UNIVERSE

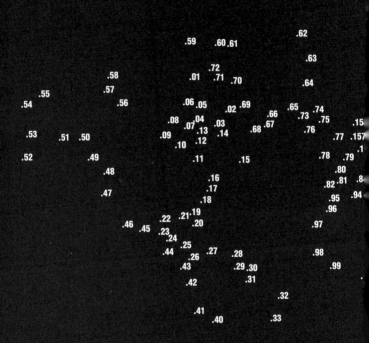

# A REALTIME UNIVERSE?

037

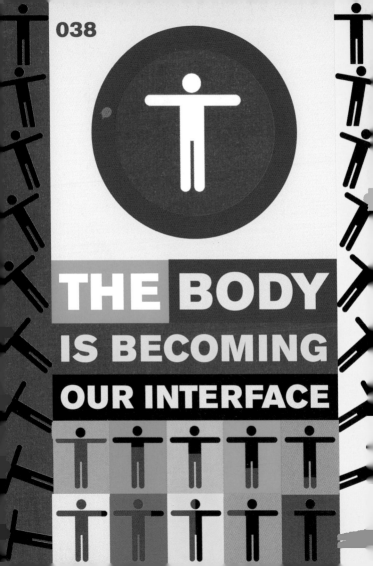

# WHEN WILL THE HARDWARE COME UP TO SPEED?

NET USERS ASSUME THAT LARGE SCREENS, COLOURFUL ICONS, POP-DOWN MENUS AND ALL THE OTHER FEATURES OF THE PC INTERFACE SHOULD BE AVAILABLE ON MOBILES WHICH ARE SMALL ENOUGH TO FIT INTO THEIR POCKETS.

# ITS NOT ABOUT BEING MOBILE,

# ITS ABOUT GETTING MOBILE

**DEVELOPMENT OF HIGH SPEED THIRD GENERATION MOBILE PHONE TECHNOLOGY has stalled THE WIRELESS APPLICATION PROTOCOL (WAP) has NOT proved a commercial success. 3G PHONES ARE PLANNED TO BE USED FOR VIDEO, an application which has already proved a flop on wired networks.**

**042**

# IS
BEING CONNECTED MORE IMPORTANT THAN **WHAT** WE'RE CONNECTING **ABOUT?**

The Internet opened up our private spaces and transformed our homes into globally shared directories. Once we withdrew from reality, we now invite reality in again. Similarly mobile connectivity increases the vulnerability of individual space with another entry for intrusion. The advantage of having availability at any time, or instant access to information, should be weighed against the disadvantages of these dependencies.

PERSONALSPACE
JUNKSPACE
VIRTUALSPACE
CELLSPACE
VISUALSPACE
FREESPACE
PUBLICSPACE
NETWORKSPACE
SOCIALSPACE
COMMUNITYSPACE
WORKSPACE
CYBERSPACE
SMARTSPACE
AUGMENTEDSPACE

> American reluctance to use mobile phones largely hinges on a highly developed sense of privacy and individuality. Just as people from more social, interconnected cultures see mobiles as a way of extending their networks and adding to their collectivity, many Americans seem to fear that the mobile will undermine their self-reliance and their independence, as well as disturbing their personal space.

**CELLSPACE NEED NOT EVEN REQUIRE A CELLPHONE**

These days, there's all kinds of

# "*beltware*"

*devices* morphing from phones or organizers or MP3 players into cellspace devices.

# MAYBE

*the trick will be to combine all these devices.*

# MAYBE

*they will become much more specific in function, so that people will carry a bunch of them, including a remote with which to find all the other beltware.*

It is not by accident that the cellphone has really taken off in places that are more urban than suburban. Places where congestion or poverty or cultural preference lead people outdoors, into the streets, into cafes and bars and piazzas, places where the Internet hasn't always been a big hit. PERHAPS THE INTERNET WAS TOO ALWAYS A BIT SUBURBAN FOR THE REST OF THE WORLD.

When the East Timorese were fighting for independence from Indonesia, Timorese leader Jose Ramos Horta commented that modems had become more important than machine guns in making their point. Perhaps it is now time for the cellphone rather than the cluster bomb.

*The clean, bright landscape of the commercial wireless world is a frictionless, transcendent illusion of unimpeded, always-on access.*

**Underneath these blue skies, users are interacting with a network that is ambiguous, unreliable and rooted firmly in a physical environment,** yet it is precisely these qualities that help us incorporate its potential into our equally ambiguous, unreliable and physical lives.

The way users navigate urban space could be crudely summarised by the opposition of two archetypes -

# THE COMMUTER
# THE FLANEUR

THE COMMUTER'S JOURNEY IS CHARACTERISED BY PATTERN, ROUTINE AND ORDER;

THE FLANEUR'S BY CAPRICIOUSNESS, UNPREDICTABILITY AND A CONCERTED EFFORT TO FLOUT THE CONVENTION OF PUBLIC SPACE.

**BEEP, BEEP.....BEEP, BEEP.....BEEP, BEEP.....BEEP, BEEP....**
OH-oh.....OH-oh.....OH-oh.....OH-oh.....OH-oh.....OH-oh.....O
PRrrr.....PRrrr.....PRrrr.....PRrrr.....PRrrr.....PRrrr.....PRrrr.....
TING, TING.....TING, TING.....TING, TING.....TING, TING.....
HAha.....HAha.....HAha.....HAha.....HAha.....HAha.....HAha

**BEEP, BEEP.....BEEP, BEEP.....BEEP, BEEP.....BEEP, BEEP..**
OH-oh.....OH-oh.....OH-oh.....OH-oh.....OH-oh.....OH-oh.....
PRrrr.....PRrrr.....PRrrr.....PRrrr.....PRrrr.....PRrrr.....PRrrr.....
TING, TING.....TING, TING.....TING, TING.....TING, TING.....
HAha.....HAha.....HAha.....HAha.....HAha.....HAha.....HAha

**BEEP, BEEP.....BEEP, BEEP.....BEEP, BEEP.....BEEP, BEEP..**
OH-oh.....OH-oh.....OH-oh.....OH-oh.....OH-oh.....OH-oh.....
PRrrr.....PRrrr.....PRrrr.....PRrrr.....PRrrr.....PRrrr.....PRrrr.....
TING, TING.....TING, TING.....TING, TING.....TING, TING.....
HAha.....HAha.....HAha.....HAha.....HAha.....HAha.....HAha

**BEEP, BEEP.....BEEP, BEEP.....BEEP, BEEP.....BEEP, BEEP..**
OH-oh.....OH-oh.....OH-oh.....OH-oh.....OH-oh.....OH-oh.....
PRrrr.....PRrrr.....PRrrr.....PRrrr.....PRrrr.....PRrrr.....PRrrr.....
TING, TING.....TING, TING.....TING, TING.....TING, TING.....
HAha.....HAha.....HAha.....HAha.....HAha.....HAha.....HAha

**BEEP, BEEP.....BEEP, BEEP.....BEEP, BEEP.....BEEP, BEEP..**
OH-oh.....OH-oh.....OH-oh.....OH-oh.....OH-oh.....OH-oh.....
PRrrr.....PRrrr.....PRrrr.....PRrrr.....PRrrr.....PRrrr.....PRrrr.....
TING, TING.....TING, TING.....TING, TING.....TING, TING.....
HAha.....HAha.....HAha.....HAha.....HAha.....HAha.....HAha

**BEEP, BEEP.....BEEP, BEEP.....BEEP, BEEP.....BEEP, BEEP.**
OH-oh.....OH-oh.....OH-oh.....OH-oh.....OH-oh.....OH-oh.....
PRrrr.....PRrrr.....PRrrr.....PRrrr.....PRrrr.....PRrrr.....PRrrr.....
TING, TING.....TING, TING.....TING, TING.....TING, TING.....
HAha.....HAha.....HAha.....HAha.....HAha.....HAha.....HAha

# THE  "beep beep"

OF AN INCOMING MESSAGE IS A REFRAIN THAT IS CAPABLE OF BRINGING HOME AND COMMUNITY INTO AN UNFAMILIAR SPACE.

 TEXT MESSAGING ALLOWS PRIVATE COMMUNICATION TO OCCUR IN PUBLIC PLACES.

This reterritorialises the spaces that users operate in, and affects their sense of self and of home.

VOICE CALLS ON MOBILE PHONES CAN BE LIMITED IN SOME SITUATIONS BECAUSE OF THE NEED FOR PRIVACY, BUT BECAUSE TEXT MESSAGING IS SILENT AND DISCREET IT CAN BE USED ANYWHERE.

# BUSINESS WEEK ABOUT PALM V DESIGNER DENNIS BOYLE:

His mission: To make high-tech simple. "People don't want to read a manual. They don't want something confusing that makes them look dumb," he says. "What regular people want is a product that does a few things really well."

Just as consumers now change screensavers and buy iMac computers in a range of colors, in the future they'll have more options when it comes to the look, feel, and functionality of high-tech products. Boyle compares it to buying a car: "Some people want a car with good gas mileage. Some want a sports car to look cool. Products have to be configured to the person using them."

For insight into future tech users, Boyle often turns to his two young sons, aged 6 and 12. "Kids want to know why things are the way they are," he notes. "Why doesn't the window in the back of the car go all the way down? Why do batteries run out so often? Their complaints give a lot of insight into what the problems are," he muses.

**Mobile phones enable the type of (virtual) communication and interaction which characterizes premodernity:**

**people who never move far, live in small towns and villages near each other, everybody knows where everybody is etc.**

But being virtual, this kind of communication is not bounding more to any single locality, as it was in premodern times. This makes it a very postmodern phenomenon.

Mobile phones make it possible to empirically test some of the claims made by postmodern authors. And that is defenitly why the subject has been so meticulously avoided.

**J. P. Roos, Postmodernity and Mobile Communications (http://www.valt.helsinki.fi/staff/jproos/mobilezation.htm)**

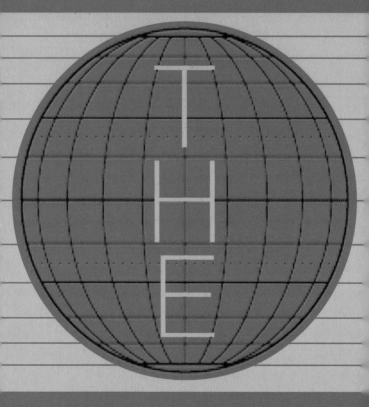

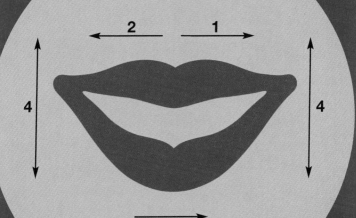

# e mobil
# le mobi
# ile mob
# bile mo
# obi lem

*MILBI TOY*

# FREEDOM
# CREATIVITY
# MOBILITY
became ordinary marketing tools
# TO BUY
# CONNECTION
# FACILITIES
# ACCOMODATION

# DISCONNECTION
## FOR LIFE
# FINE
## FOR CELLPHONE
# ABUSE

# ST HELENS COLLEGE
## Self Service Station

Olivia Maguire

LOAN  22/01/2019 12:15:53

1. 110365  Check-in: **25/02/2019**
   Mobile minded

2. 124320  Check-in: **25/02/2019**
   Graphics

**MOBILITY IS THE MENTAL TERRITORIAL MAP OF YOUR LIFE**

*connectivity is the database of THE RELATIONSHIPS you move around with*

# Freedom: mobility unplugged

065

**IN GENERAL:**
- 'GOOD TASTE IS THE REFUGE OF THE WITLESS.'
  HARLEY PARKER
  OPENING QUOTE IN COUNTER BLAST,
  MARSHALL MCLUHAN, 1969
- 'THE URB IT ORBS.'
  JAMES JOYCE

**ON OUR ELECTRONIC INFORMATION ENVIRONMENT:**
- ENVIRONMENT IS PROCESS, NOT CONTAINER.
- WE LIVE IN AN ELECTRONIC INFORMATION ENVIRONMENT THAT IS QUITE AS IMPERCEPTIBLE TO US AS WATER IS TO A FISH.
- CONDITIONING DEPENDS ON A PREVIOUS CONDITIONING.
- OUR ILLUSION OF 'CONTENT' DERIVES FROM ONE MEDIUM BEING 'WITHIN' OR SIMULTANEOUS WITH ANOTHER.
- WHY PROGRAM THE ENVIRONMENT INSTEAD OF ITS CONTENT?

**ON SOCIAL ENACTMENT:**
- THRIVE ON THE TERRAIN WHERE HUMAN HAND NEVER SET FOOT!
- POWER TO THE INTERMIDGETS!
- LOVE THY LABEL AS THYSELF.

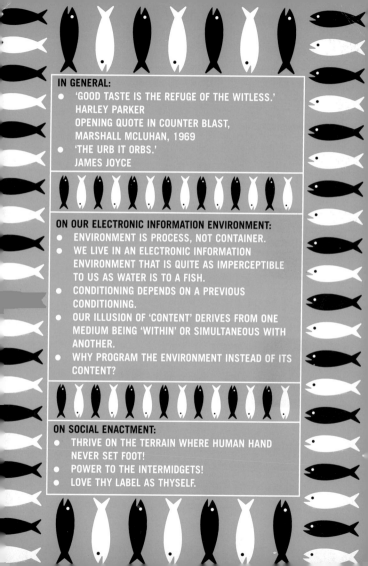

*The mobile device as* **creative platform** *could prove to be a fascinating and perhaps slightly threatening* place.

BILINGUAL

HYPER TALK

ハイハートーク
対応

Convergence of PDA, telephony, gaming, and musical technologies **into mobiles** offer the possibility

ハイフレータ

of simultaneous modes of expression and collective **creation not seen before.**

# DON'T SACRIFICE THE ABILITY TO UPLOAD FOR MOBILITY

*downloading IS consumption*

069

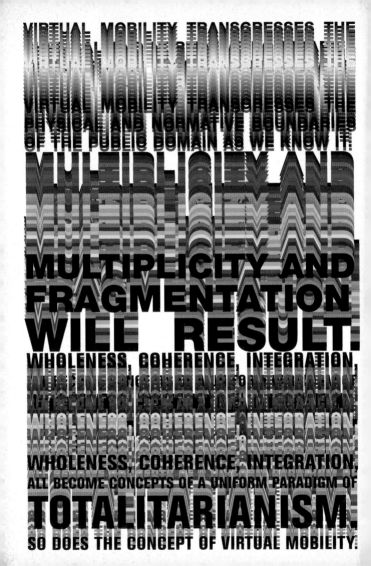

# The e-word becomes the m-word:

## ALL PATHETIC SIMPLICITY.

071

the mobile minded:

# just

# another

### tribe of

# command

### and

# control

### freaks

THE MOBILE WORLD OF VIRTUALITY/
THE VIRTUAL WORLD OF MOBILITY:
INSTEAD OF PETTY BOURGEOIS
DREAMS OF THE SIXTIES,
NEW RHIZOMATIC STRATEGIES
AND CONFIGURATIONS OF POWER.

# the city
## of tomorrow:

**AMSTERDAM**
**NEW YORK**
**LISBOA**
**PARIS**
**BARCELONA**
**LONDON**
**COPENHAGEN**
**MELBOURNE**
**BERLIN**
**MILANO**
**MOSKOW**

**SHOULD WE SUBSIDIZE MOBILE SERVICES FOR THE "HANDICAPPED", LIKE WE HAVE LAWS THAT GOVERN ACCESSIBILITY TO BUILDINGS?**

**10:00 AM**
PARIS CDG AIRPORT
**WAITING AREA**
**COMFORTABLE**
WHITE LEATHER CHAIRS
120 BEEPS PER MINUTE
**I SWITCH OFF**
**ALL DEVICES**
**TO GET READY**
**FOR BOARDING**

SOMEBODY TAPS ON MY SHOULDER POINTING TO THE CHAIR I WAS SITTING ON. "EXCUSE ME, IS THAT YOUR MIND YOU LEFT OVER THERE?"

**WHEREVER YOU GO YOU WILL BE ABLE TO TAKE PART.**

**WHEREVER YOU GO YOU WILL BE ABLE TO STAY INFORMED.**

**WHEREVER YOU GO YOU WILL BE ABLE TO HAVE AN OPINION.**

**WHEREVER YOU GO YOU WILL BE ABLE TO SAY YOUR MEANING.**

**WHEREVER YOU GO YOU WILL BE RESPONSIBLE.**

It is important not to loose sight of the flashes of mobility through the forest of cel antennas. Mobility is not cellphones, airplanes, cars or instant messaging. It is also worth noting that we are by our nature mobile. Mobility is how we live. It is our movement from birth to death, the transformations of our body and our mind as we grow older. It is the quest that propels us, that lies at the heart of all literature. **There is no absolute state of mobility, nor one of stasis.** My memory of great things includes a catalogue of landscapes glimpsed from car windows and northern lights seen from airplanes, of words heard in other places. **Mobility was not invented by Nokia. It did not suddenly come into our lives with the combustion engine, nor does it disappear when we press ctrl-alt-del. The gadgets that increase** our mobility are only that: tools that enhance our ability to experience and that can make our path through life infinitely more complex. **It is true that more and more of our time is spent moving without meaning, and that always the pressure to keep moving is at our back --but then so are "time's winged chariots."** It is what we make in the wave of that pressure that makes us real, that reminds us what body it is that is moving along the tides of history, to quote yet another cliche of movement. **What do I want from mobility?** To surf. What I treasure about mobility are the moments of stillness suspended thirty thousand feet above the air, the voice of my loved one as I sit in an anonymous conference center in Madrid, the architecture I love because it catches all that motion in one fact.

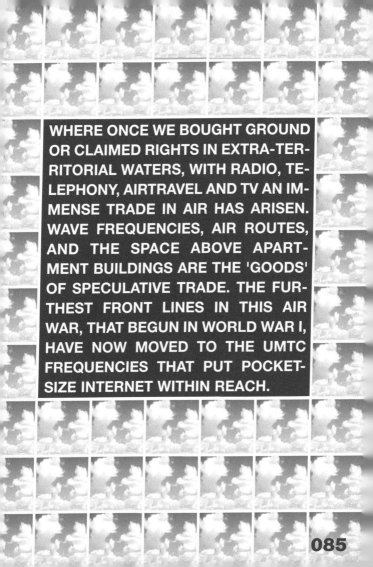

WHERE ONCE WE BOUGHT GROUND OR CLAIMED RIGHTS IN EXTRA-TERRITORIAL WATERS, WITH RADIO, TELEPHONY, AIRTRAVEL AND TV AN IMMENSE TRADE IN AIR HAS ARISEN. WAVE FREQUENCIES, AIR ROUTES, AND THE SPACE ABOVE APARTMENT BUILDINGS ARE THE 'GOODS' OF SPECULATIVE TRADE. THE FURTHEST FRONT LINES IN THIS AIR WAR, THAT BEGUN IN WORLD WAR I, HAVE NOW MOVED TO THE UMTC FREQUENCIES THAT PUT POCKET-SIZE INTERNET WITHIN REACH.

I consider radical mediocrity as a quality of the way we deal with all kinds of media.

A MEANS CAN BE THE SUBJECT OR OBJECT OF THE MEASURE: it can indicate the measure or be measured itself. The measure in turn can be interpreted in just as many ways: it manifests itself sometimes as movement, sometimes as measurement.

HOW MUCH A MEANS - PROZAC, COCAINE, CAR, MOBILE PHONE - IS USED IS NEVER A CONSIDERED CHOICE BY INDIVIDUALS, BUT A SITUATIONALLY DETERMINED ROUTINE.

# Is there such a thing as NETiquette for mobile use ?

'CONNECTIVITY' is a smart marketing promise, luring us into the illusion of independence and flexibility. Obviously, instant and location independent access to communication networks has its many advantages. It provides us with the option to act or respond instantaneously, any time, anywhere.

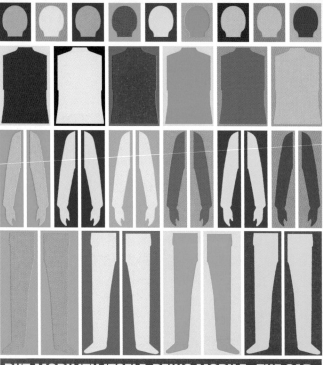

BUT MOBILITY ITSELF, BEING MOBILE - THE CAPABILITY OF MOVING - SHOULD NOT BE COMPARED AND JUDGED BY THE SAME SET OF VALUES.

**CARVING OUT MOBILE SPACE IS GOOD, using it to RECLAIM PUBLIC SPACE is better.**

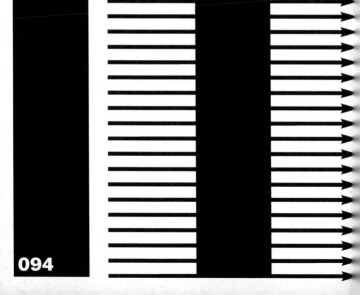

# IN ADS THEY SAY:

*"surf the web on your phone"*

# WILL IT EVER HAPPEN THAT YOU WILL SEE ADS SAYING:

*"surf the mobileweb via the NET"*

# THAT WAY YOU PAY TWICE.

# The unmoved mover.

Seated in front of our golden calf, the shining screen, we let the world come to us, and it shoots past us, located inside the sacred cow. With the merest touch of the remote control, mouse or gearshift the world turns or is brought to a standstill, as we wish.

# This immobilized individual is nevertheless more mobile than mobile: it is in a **PANIC.**

This panic manifests itself chiefly through the alert maintenance of all states of readiness: at any moment anywhere accessible and available. So we run around aimlessly, without moving, in the atopia of speed, a no man's land at the furthest boundary of our history, where freedom of movement has become compulsory and a collective desire for consumption seems to be the only reference point to cross time and space.

# ARE WE FRYING OUR BRAINS INSTEAD OF POLLUTING OUR LUNGS?

For today's young people, the first sign of maturity is ignoring the danger of radiation from mobiles rather than disregarding the risk of getting cancer from smoking.

**MOBILE DEVICES FROM THE RENAISSANCE.** the left and right wing of an altar painted by Francesco Carotto. saint lucie (saint of the blind) is grabbing something with her "lily-camera" sending it to saint george who is receiving this with his "roses monitor".

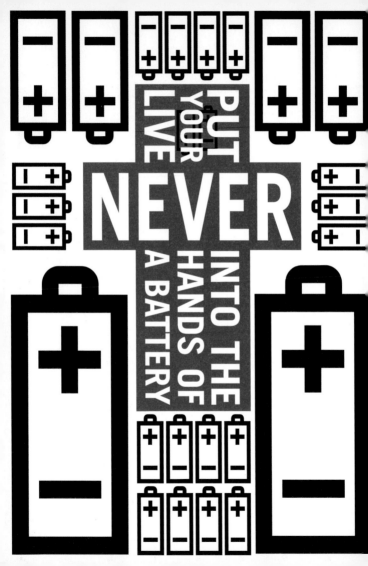

ENJOYING BY SHARING WALKING ALONG THE AUSTRALIAN BEACH I CALLED THE BOY I WANTED TO SEDUCE TO LET HIM HEAR THE SOUNDS OF THE BYRON BAY OCEAN WAVES. WALKING THROUGH AN AUTUMN FOREST I PHONED MY FRIEND TO ENSURE THAT THE FUNGI I PICKED WAS AN EDIBLE ONE. AND ON THE TOP OF A DISTANT MOUNTAIN IT ANNOYED ME TERRIBLY THAT I WAS UNABLE TO CALL ANYONE TO SHARE THE SONG OF A MOUNTAIN BIRD, AS THERE WAS NO MOBILE NETWORK.

> Perversely, mobility seems to reinforce the importance of the physical location of the user - most people's mobile conversations often start with the question:

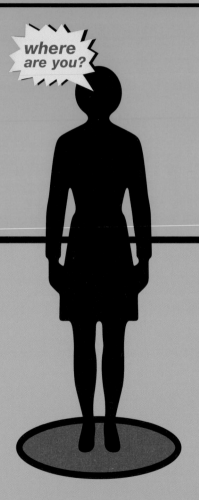

*where are you?*

Mobile networks have to negotiate the architecture of spaces that they attempt to inhabit. Although the interfaces have removed themselves from physical architectures, the radio waves that connect cell spaces are refracted and reflected by these same obstacles, creating not a seamless network but a series of ebbs and flows. The supposedly flat surface of the network is in fact warped, pulled into troughs and peaks by the gravity of architecture and the users themselves.

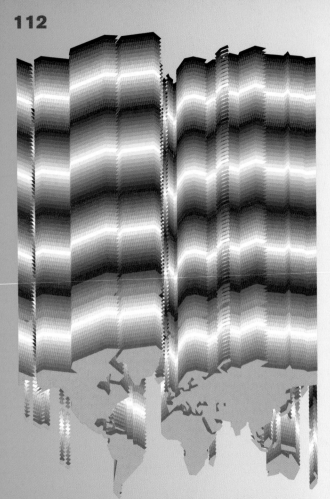

MOBILITY IS NOT ONLY ABOUT BEING ACCESSIBLE TO EVERYONE AND EVERYTHING AT ANYTIME, IT'S ALSO ABOUT BEING AWARE OF OTHERS CONNECTED TO YOU. MOBILITY IS ALSO ABOUT HOW WE CAN USE TECHNOLOGY TO INTEGRATE AND LINK OUR EXPERIENCES TO OTHERS. IN THE REAL WORLD WHEN WE WALK AROUND WE ARE PHYSICAL BROWSERS TAKING IN DIFFERENT INFORMATION IN THE FORM OF SIGNS, STORES, LOCATIONS, PEOPLE, ETC. WITH TECHNOLOGY WE CAN CAPTURE OUR EXPERIENCES IN THE WORLD AND CREATE A VAST DATABASE OF REAL-TIME LOCATION-BASED INFORMATION THAT CAN BE DYNAMICALLY RELAYED TO PEOPLE CONNECTED TO THE NETWORK. IN THIS SENSE, PEOPLE THEMSELVES CREATE AN AMORPHOUS, ORGANIC NETWORK AND INSTEAD OF EVERYONE CONNECTING TO A CENTRAL CELL-BASED SERVICE THEY CONNECT TO EACH OTHER DIRECTLY.

GET LOST

# MOBILE MINDED

# OPEN DIALOGUES

**THE MOBILE MIND** *THIS SEEMINGLY OLD* **PARADOXICALLY** *FASHIONED IDEA WILL* **ONLY WORKS** *EVEN LAST IN THE FUTURE* **ON THE BASIS** *BECAUSE HUMAN BEINGS* **OF CLINGING TO** *NEED THE CONFIRMATION* **THE LOCALIZATION** *OF THEIR EMPIRICAL LIVE-* **OF A CONCRETE PLACE** *LINESS IN PLACE IN ORDER* **TO MAKE SURE** *TO DESIGN INNOVATIVE* **THAT THE INDIVIDUAL** *AND IMAGINATIVE VIRTUAL* **(ONESELF)** *COMMUNICATION.* **"IS STILL THERE".**

# INFORMATION

# AS

# DECORATION

# MORE MOBILITY

# =

# LESS
# PRIVACY

The mobile device is another step towards computational ubiquity, as Digital devices begin to colonize the body, and will soon translate into affordable wearable computers, which may allow the creation of 'smart space'. Undoubtedly, there will be issues as to the nature of privacy and individualism in these environments, but these are merely extensions of questions being raised by the Internet right now.

# THE SILENT MAJORITY (Eric Taub, The New York Times)

**CELL YELL HAS CREATED A SUBCULTURE OF CELL-YELL HATERS**

"MOBILE PHONES ARE SO SMALL THAT PEOPLE DON'T TRUST THE TECHNOLOGY TO WORK" said Timo Kopomaa, a social scientist at the University of Technology in Helsinki and author of a study on mobile phone behaviour.

Because the mouth piece of the typical mobile phone barely extends to the cheek, many users act, consciously or not, as if they have TO SHOUT to be heard.

USERS ALSO OFTEN ENGAGED IN "STAGE-PHONING"; THAT IS, MAKING UNIMPORTANT CALLS IN PUBLIC JUST TO IMPRESS OTHERS.

Rather, a rapid response to a ring showed bystanders that the users had "TELECREDIBILITY", Kopomaa said. They had mastered this new technology, and they did not have to fumble to figure out how to answer it

MOBILITY AND PRIVACY ARE INVERSELY PROPORTIONAL QUANTITIES

# MOBILITY DISABILITY

*Mobile networks assume mobile bodies but we need to design for a heterogeneously able society. Our capacity for freedom is dependent upon the built and virtual environments of our public spheres. How might our designs change if we stopped locating liberty in individual bodies?*

```
PUBLIC CLASS INTHESTATEOFMOBILITY

{public static void main(String[ ] args)

{char mobility = {'poetic', 'organic', 'historic',
'spatial', 'dangerous', 'existential', 'situated',
'meaning-making', 'perceived', 'systemic',
'invisible', 'virulent', 'local', 'ambulant', 'con-
trolled', 'tender', 'explorative', 'blasted',
'trustworthy', 'social', 'distorted', 'fast', 'inter-
generational', 'stretched', 'affective',
'prosthetic', 'inner', 'prosthetic', 'disturbing',
'public', 'dystopian', 'olfactory', 'augmented',
'silent', 'cultural', 'wide', 'present', 'agent',
'permeable', 'erratic', 'trafficked', 'cosmic',
'mapped', 'psychic', 'functional', 'terroristic',
'rounded', 'sexual', 'global', 'vertical', 'distrib-
uted', 'gendered', 'architectural', 'rhythmic',
'collaborative', 'critical', 'intense', 'thinking',
'everydaypractical', 'individual', 'informatio-
nal', 'disabled', 'virtual', 'immersive',
'emissive', 'resistant', 'warlike'} ;

boolean[ ] dimensions = {true, nor, false} ;

System.out.println("Wherever you are,
"+mobility.random+" mobility equals ") ;

System.out.println(dimensions.random) ;
```

# NO ONE IS EVER MORE THAN TEMPORARILY MOBILE

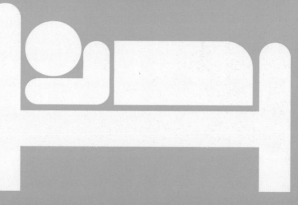

# MORE MOBILE
# =
# MORE LIBERAL

**California:**
smoking prohibited in bars, restaurants and public spaces, cellphones allowed if not stated otherwise.

**Holland:**
cellphones prohibited in bars and restaurants, smoking allowed no matter what. Until death do us part.

# DRIVE BY DINING

## WE ARE WIRELESS. WE ARE WATCHED.

SERVICE STATION — FISH

SERVICE STATION — VEGG

SERVICE STATION — MEAT

# CONTENTSPACE
# NETWORKSPACE

| 002 | Geert Lovink / Mieke Gerritzen |
|-----|-------------------------------|
| 004 | Mc Kenzie Wark |
| 006 | Manuela Porceddu |
| 007 | Mc Kenzie Wark |
|     | Image: NTT Docomo |
| 008 | Kevin Kelly |
| 009 | Michael Erlhoff |
| 010 | Matthew Fuller |
| 011 | Max Bruinsma |
| 012 | Jan van den Berg |
| 013 | Martijn Hazelzet |
| 014 | Steve Heller |
| 015 | Martijn Hazelzet |
| 016 | Henk Oosterling |
| 017 | Richard Barbrook, Andrew |
|     | Purdy, Armin Medosch, Mark |
|     | Fitzpatrick, Niki Gomez, Robin |
|     | Hamman, Sophia Drakopoulou |
| 018 | Richard Barbrook, Andrew |
|     | Purdy, Armin Medosch, Mark |
|     | Fitzpatrick, Niki Gomez, Robin |
|     | Hamman, Sophia Drakopoulou |
| 020 | Martijn Hazelzet |
| 021 | McKenzie Wark |
| 022 | Michael Erlhoff |
| 023 | Erik Spiekerman |
| 024 | Mc Kenzie Wark |
| 026 | Aaf van Essen |
| 027 | Stefan Ytterborn |
| 028 | Richard Barbrook, Andrew |

|     |                               |     |               |
| --- | --------------------------------- | --- | ----------------- |
|     | Purdy, Armin Medosch, Mark        | 051 | Matt Locke        |
|     | Fitzpatrick, Niki Gomez, Robin    | 052 | Suzy Small        |
|     | Hamman, Sophia Drakopoulou        | 054 | Dennis Boyle      |
| 029 | Stefan Sagmeister                 | 055 | J.P.Roos          |
|     | Image: Laura Kurgan               | 056 | Max Kisman        |
| 030 | Sigga Sigurjonsdottir             | 057 | Max Kisman        |
|     | Image: NTT Docomo                 | 058 | Max Kisman        |
| 031 | Douglas Ruskoff                   | 059 | Michael Erlhoff   |
| 032 | Martijn Hazelzet                  | 060 | Paul Mijksenaar   |
| 033 | Kevin Kelly                       | 061 | Matt Locke        |
| 034 | Max Kisman                        | 062 | Max Kisman        |
| 035 | Martijn Hazelzet                  | 063 | Martijn Hazelzet  |
| 036 | Martijn Hazelzet                  | 064 | Max Kisman        |
|     | Image: Silke Wawro                | 065 | Max Kisman        |
| 038 | Sigga Sigurjonsdottir             | 066 | Willem van Weelden|
| 039 | Richard Barbrook, Andrew          | 068 | Patrick Lichty    |
|     | Purdy, Armin Medosch, Mark        | 069 | Peter Lunenfeld   |
|     | Fitzpatrick, Niki Gomez, Robin    | 070 | Paul Frissen      |
|     | Hamman, Sophia Drakopoulou        | 071 | Paul Frissen      |
| 040 | Tom Worthington                   | 072 | Paul Frissen      |
| 041 | Aaf van Essen                     | 073 | Paul Frissen      |
| 042 | Peter Lunenfeld                   | 074 | Volker Albus      |
| 043 | Max Kisman                        | 075 | Richard Joly      |
| 044 | Sadie Plant                       | 076 | Peter Lunenfeld   |
| 045 | Lev Manovich                      | 077 | Lies Ros          |
| 046 | McKenzie Wark                     | 078 | Susanna Dulkinys  |
| 047 | McKenzie Wark                     | 079 | Luna Maurer       |
| 048 | Rob Schröder                      | 080 | Björn Dahlström   |
| 049 | McKenzie Wark                     | 081 | Geert Lovink      |
| 050 | Matt Locke                        | 082 | Aaron Betsky      |

| | |
|---|---|
| 084 | Henk Oosterling |
| 085 | Henk Oosterling |
| 086 | Henk Oosterling |
| 087 | Patrick Lichty |
| 088 | Image: Natacha Vairo |
| 089 | Richard Joly |
| 090 | Peter Lunenfeld |
| 091 | Richard Joly |
| 092 | Max Kisman |
| 093 | Peter Lunenfeld |
| 094 | Jan van den Berg |
| 095 | Image: Tara Karpinski |
| 096 | Richard Joly |
| 097 | Andreu Balius |
| 098 | Henk Oosterling |
| 099 | Henk Oosterling |
| 100 | Paul Mijksenaar |
| 101 | Richard Barbrook, Andrew Purdy, Armin Medosch, Mark Fitzpatrick, Niki Gomez, Robin Hamman, Sophia Drakopoulou |
| 102 | Joachim sauter |
| 104 | Martijn Hazelzet |
| 105 | Annelys de Vet |
| 106 | Matt Locke |
| 107 | Silke Wawro |
| 111 | Matt Locke |
| 112 | Andreu Balius |
| 113 | Sigga Sigurjonsdottir |
| 114 | Jonah Brucker-Cohen |
| 115 | Jan van den Berg |
| 116 | Andreu Balius |
| 117 | Uta Brandes |
| 118 | Jan van den Berg |
| 119 | Mieke Gerritzen |
| 120 | Marcel Vosse |
| 121 | Patrick Lichty |
| 122 | Eric Taub |
| 123 | Eric Kluitenberg |
| 124 | Marcel Vosse |
| 125 | Meg McLagan |
| 126 | Steffen P Walz |
| 127 | Meg McLagan |
| 128 | Rob Schröder |
| 129 | Max Kisman |
| 130 | Andreu Balius |
| 131 | Shulea Cheang |

# PEOPLE

| | | |
|---|---|---|
| DE | **VOLKER ALBUS** | |
| | PROFESSOR PRODUCT DESIGN, FRANKFURT | |
| ES | **ANDREU BALIUS** | |
| | DESIGNER, BARCELONA | |
| UK | **RICHARD BARBROOK,** | |
| | COORDINATOR HYPERMEDIA RESEARCH CENTRE, WESTMINSTER UNIVERSITY | |
| NL | **JAN VAN DE BERG** | |
| | DESIGNER, AMSTERDAM | |
| DE | **UTA BRANDES** | |
| | PROFESSOR GENDER DESIGN, COLOGNE | |
| US | **AARON BETSKY** | |
| | DIRECTOR NETHERLANDS ARCHITECTURE INSTITUTE, ROTTERDAM | |
| US | **STEWART MCBRIDE** | |
| | DIRECTOR UNITED DIGITAL ARTISTS, NEW YORK/PARIS | |
| US | **JONAH BRUCKER-COHEN** | |
| | ARTIST / FELLOW MIT DUBLIN | |
| NL | **MAX BRUINSMA** | |
| | EDITOR, AMSTERDAM | |
| US | **SHULEA CHEANG** | |
| | DIGITAL ARTIST, GLOBAL | |
| SE | **BJÖRN DAHLSTRÖM** | |
| | INDUSTRIAL DESIGNER, STOCKHOLM | |
| US | **SUSANNA DULKINYS** | |
| | DESIGNER, SAN FRANCISCO | |
| DE | **MICHAEL ERLHOFF** | |
| | PRESIDENT RAIMOND LOEWY FOUNDATION, COLOGNE | |
| NL | **AAF VAN ESSEN** | |
| | HEAD SANDBERG INSTITUTE (DESIGN), AMSTERDAM | |
| NL | **PAUL FRISSEN** | |
| | PROFESSOR OF PUBLIC ADMINISTRATION, KUB, TILBURG | |

| | | |
|---|---|---|
| NL | MIEKE GERRITZEN | |
| | FOUNDER AND DESIGNER OF NL.DESIGN, AMSTERDAM | |
| NL | MARTIJN HAZELZET | |
| | DIRECTOR ORGNET, DEN HAAG | |
| US | STEVE HELLER | |
| | EDITOR, NEW YORK TIMES | |
| CAN | RICHARD JOLY | |
| | INTERNET RESEARCH & BUSINESS, MONTREAL | |
| US | TARA KARPINSKI | |
| | DESIGNER, AMSTERDAM | |
| US | KEVIN KELLY | |
| | EDITOR AT LARGE WIRED AND WRITER, SAN FRANCISCO | |
| NL | MAX KISMAN | |
| | DESIGNER, SAN FRANCISCO | |
| NL | ERIC KLUITENBERG | |
| | PROGRAMMER MEDIA EVENTS, DE BALIE, AMSTERDAM | |
| US | PATRICK LICHTY | |
| | TECHNOLOGICALLY BASED CONCEPTUAL ARTIST, USA | |
| UK | MATT LOCKE, CREATIVE DIRECTOR AT BBC ENGINEERING, FORMERLY AT THE | |
| | MEDIA CENTRE, HUDDERSFIELD | |
| NL | GEERT LOVINK | |
| | MEDIA THEORIST AND INTERNET CRITIC, AUSTRALIA | |
| US | PETER LUNENFELD, DIRECTOR, GRADUATE FACULTY COMMUNICATION | |
| | AND NEW MEDIA DESIGN, ART CENTER COLLEGE OF DESIGN, PASADENA | |
| US | LEV MANOVICH | |
| | NEW MEDIA ARTIST AND THEORIST, UNIVERSITY OF CALIFORNIA, SAN DIEGO | |
| DE | LUNA MAURER | |
| | DESIGNER, AMSTERDAM | |
| US | MEG MCLAGAN | |
| | INTERNATIONAL INSTITUTE FOR ASIAN STUDIES, AMSTERDAM/ NYU, NEW YORK | |
| NL | PAUL MIJKSENAAR | |
| | DIRECTOR BUREAU MIJKSENAAR, AMSTERDAM | |
| NL | HENK OOSTERLING | |
| | PROFESSOR AND PHILOSOPHER, ROTTERDAM | |
| UK | SADIE PLANT | |
| | WRITER AND CRITIC, LONDON | |

| | | |
|---|---|---|
| NL | MANUELA PORCEDDU | |
| | DESIGNER, ROTTERDAM | |
| NL | LIES ROS | |
| | DESIGNER, AMSTERDAM | |
| US | DOUGLAS RUSKOFF | |
| | PROFESSOR AND WRITER, NEW YORK | |
| NL | MARJOLIJN RUYG | |
| | DESIGNER, AMSTERDAM | |
| US | STEFAN SAGMEISTER | |
| | DESIGNER, NEW YORK | |
| DE | JOACHIM SAUTER | |
| | DESIGNER, ART&COM, BERLIN | |
| NL | ROB SCHRODER | |
| | VPRO-TELEVISION, HILVERSUM | |
| IS | SIGRIDUR SIGURJONSDOTTIR | |
| | DESIGNER, LONDON | |
| AU | SUZY SMALL | |
| | MOBILE PHONE RESEARCHER, UNIVERSITY OF NEW SOUTH WALES | |
| DE | ERIK SPIEKERMANN | |
| | DESIGNER, BERLIN/SAN FRANCISCO | |
| CAN | NATACHA VAIRO | |
| | DESIGNER, AMSTERDAM | |
| NL | ANNELYS DE VET | |
| | DESIGNER, AMSTERDAM | |
| NL | MARCEL VOSSE | |
| | COPYWRITER LOWELIVE, AMSTERDAM | |
| US | STEFFEN P WALZ | |
| | SPACE TIME CONSULTANT, STUTTGART | |
| AU | MCKENZIE WARK | |
| | MEDIA THEORIST AND CRITIC, USA | |
| DE | SILKE WAWRO | |
| | DESIGNER, AMSTERDAM | |
| NL | WILLEM VAN WEELDEN | |
| | PROFESSOR HKU/RIETVELD ACADEMY, HILVERSUM/AMSTERDAM | |
| SE | STEFAN YTTERBORN | |
| | DIRECTOR YTTERBORN & FUENTES, STOCKHOLM | |

# OTHER NL.DESIGN PUBLICATION:

# EVERYONE IS A DESIGNER! MANIFEST /////////FOR THE///////// DESIGN ECONOMY

Everyone is a designer: 100 tips for a design economy.
This collection of slogans, comments, maxims and glossaries is meant to boost your digital creativity. Rich in magic formulas, this publication will enhance your global network of sources for inspiration.

140 pages, paperback, 105 x 150 mm, full color, English

Published by BIS Publishers. Co-publisher: Gingko Press (USA)

For information about sales in your area, please contact BIS Publishers, The Netherlands, T +31 (020) 6205171, E-mail: bis@bispublishers.nl
http://www.bispublishers.nl